I0824574
DOLPHIN
ALLIGATOR
BUFFALO
BISON
CROCODILE
MANATEE
TURTLE
TORTOISE
DUGONG
TIGER
LEOPARD
PORPOISE
APE
MONKEY
DOLPHIN
ALLIGATOR
BUFFALO
BISON
CROCODILE
MANATEE
TURTLE
TORTOISE
DUGONG
TIGER
TORTOISE
LEOPARD
APE

For my sibs Page, Steve, and Anne. You know why!—M. F.

For Courtney, the tiger to my leopard—S. S.

BEACH LANE BOOKS • An imprint of Simon & Schuster Children's Publishing Division • 1230 Avenue of the Americas, New York, New York 10020 • For more than 100 years, Simon & Schuster has championed authors and the stories they create. By respecting the copyright of an author's intellectual property, you enable Simon & Schuster and the author to continue publishing exceptional books for years to come. We thank you for supporting the author's copyright by purchasing an authorized edition of this book. • No amount of this book may be reproduced or stored in any format, nor may it be uploaded to any website, database, language-learning model, or other repository, retrieval, or artificial intelligence system without express permission. All rights reserved. Inquiries may be directed to Simon & Schuster, 1230 Avenue of the Americas, New York, NY 10020 or permissions@simonandschuster.com. • Text © 2026 by Meg Fleming • Illustration © 2026 by Steph Stilwell • Book design by Steph Stilwell and Lauren Rille • All rights reserved, including the right of reproduction in whole or in part in any form. • BEACH LANE BOOKS and colophon are trademarks of Simon & Schuster, LLC. • For information about special discounts for bulk purchases, please contact Simon & Schuster Special Sales at 1-866-506-1949 or business@simonandschuster.com. • Simon & Schuster strongly believes in freedom of expression and stands against censorship in all its forms. For more information, visit BooksBelong.com. • The Simon & Schuster Speakers Bureau can bring authors to your live event. For more information or to book an event, contact the Simon & Schuster Speakers Bureau at 1-866-248-3049 or visit our website at www.simonspeakers.com. • The text for this book was set in Sofia Pro Soft. • The illustrations for this book were rendered with ink on paper and then digitally colored in Photoshop. • Manufactured in China • 1225 SCP • First Edition • 2 4 6 8 10 9 7 5 3 1 • Library of Congress Cataloging-in-Publication Data • Names: Fleming, Meg author | Stilwell, Steph illustrator • Title: Pretty close, but not the same : a side-by-side look at confusable critters / Meg Fleming ; illustrated by Steph Stilwell. • Description: First edition. | New York : Beach Lane Books, 2026. | Audience: Ages 4–8 | Audience: Grades 2–3 | Summary: "Discover seven pairs of similar animals that are easily confused for each other, and what makes them unique!"— Provided by publisher. • Identifiers: LCCN 2025024610 (print) | LCCN 2025024611 (ebook) | ISBN 9781665978996 hardcover | ISBN 9781665979009 ebook • Subjects: LCSH: Animals—Juvenile literature | Anatomy, Comparative—Juvenile literature | CYAC: Animals | Comparative anatomy | LCGFT: Picture books • Classification: LCC QL806.5 .F57 2026 (print) | LCC QL806.5 (ebook) | DDC 590—dc23/eng/20250915 • LC record available at https://lccn.loc.gov/2025024610 • LC ebook record available at https://lccn.loc.gov/2025024611

PRETTY CLOSE, but NOT the SAME

A Side-by-Side Look at Confusable Critters

words by
MEG FLEMING

pictures by
STEPH STILWELL

BEACH LANE BOOKS
New York Amsterdam/Antwerp London Toronto Sydney/Melbourne New Delhi

Look at that

TORTOISE!

Look at that

TURTLE!

I’m not a tortoise.

I’m not a turtle!

We've got the shell. And matching beak.
My shell has bumps.
TORTOISE

Similar in shape and frame.
Pretty **close.**
But not the same.

Look at that

MONKEY!

Look at that

APE!

I'm not a monkey.
I'm not an ape!

We both love trees.
We hang. We're fun!
But I've
got a tail.
MONKEY

APE
And I've got none.
It's not your fault.
You're not to blame.
Pretty close.
But not the same.

Look at that

DOLPHIN!

Look at that

PORPOISE!

I'm not a porpoise!
I'm not a dolphin.

We smile. We play.
We squeak. We spout.
DOLPHIN
But I've got
a bottlenose.

Jumping is our claim to fame.
Pretty close.
But not the same.

Look at that

LEOPARD!

Look at that

TIGER!

I'm not a leopard!
I'm not a tiger!

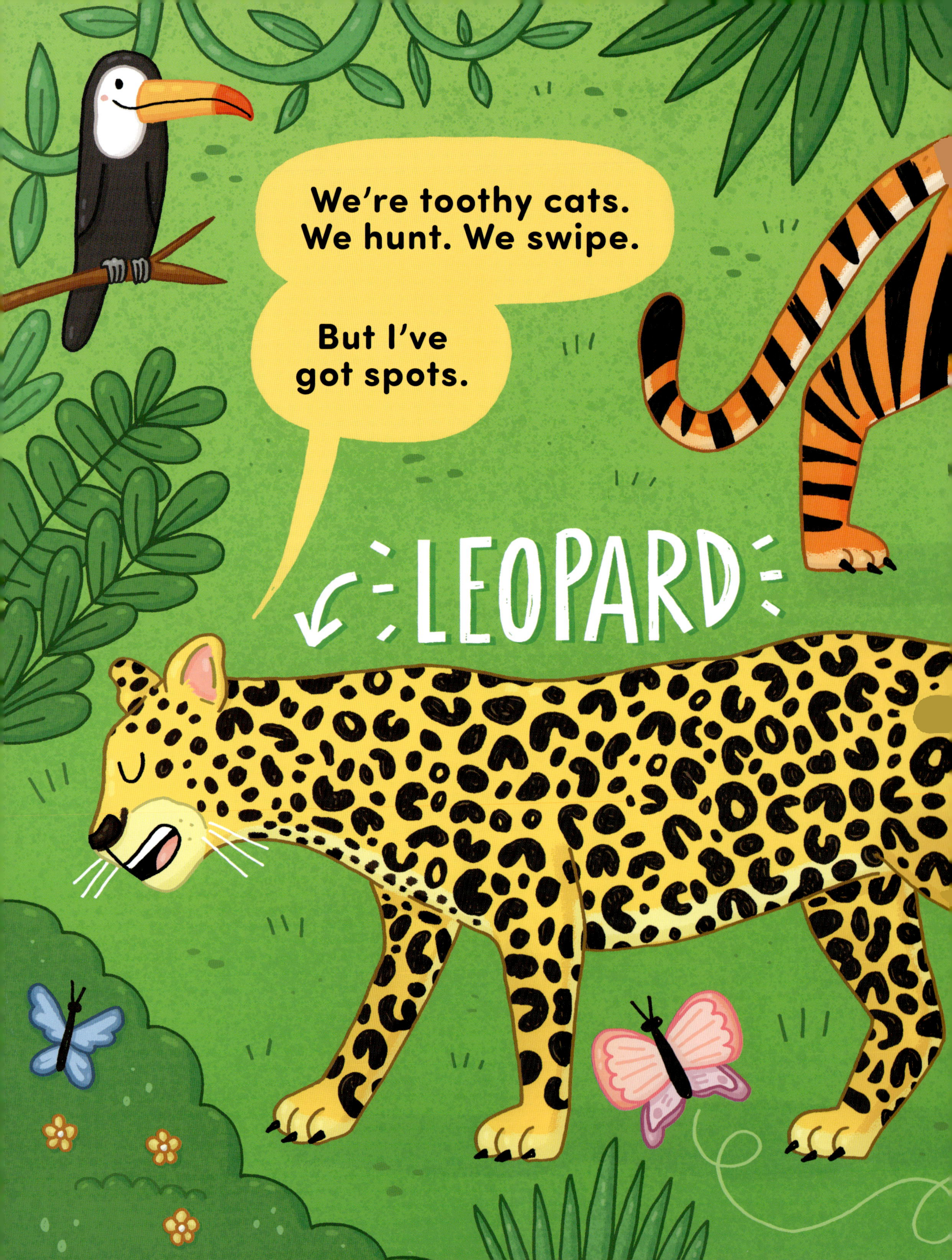
We're toothy cats.
We hunt. We swipe.
But I've
got spots.
LEOPARD

TIGER
And I've got stripes.
Mix us up?
Oh . . . we'll complain!
Pretty close.
But not the same.

Look at that

MANATEE!

Look at that

DUGONG!

I'm not a manatee.
I'm not a dugong!

We float. We sink. And what a tail!
Mine's like a paddle.
MANATEE

DUGONG
Mine's like a whale.
We're not upset.
Just fix the name.
Pretty close.
But not the same.

Look at that

CROCODILE!

Look at that

GATOR!

I'm not a crocodile.
I'm not a gator!

We watch.
We wait.
The things
we see!
ALLIGATOR
My snout's like a U.

Don't mess with us.
It's not a game.
Pretty close.
But not the same.

Look at that

BUFFALO!

Look at that

BISON!

I'm not a
buffalo.

I'm not
a bison.

We're big. We're buff.
We should be feared.
But I've got
a uni-horn.
BUFFALO

Back it up!
We're far from tame.
Pretty close.
But not the same.

So now you know.
We're glad you came.

We're
lots
alike...

BUT
NOT

THE
SAME!

DOLPHIN
ALLIGATOR
BUFFALO
BISON
CROCODILE
MANATEE
TURTLE
TORTOISE
DUGONG
TIGER
LEOPARD
APE
MONKEY
PORPOISE